Showd

Animal Groups

Heather E. Schwartz

Publishing Credits

Rachelle Cracchiolo, M.S.Ed., *Publisher*
Conni Medina, M.A.Ed., *Managing Editor*
Nika Fabienke, Ed.D., *Series Developer*
June Kikuchi, *Content Director*
John Leach, *Assistant Editor*
Kevin Pham, *Graphic Designer*

Image Credits: pp.8, 9, 16 bottom: Martin Strmiska/Alamy; pp.10, 11, 16 middle: Ivan Vdovin/Alamy; all other images from iStock and/or Shutterstock.

Library of Congress Cataloging-in-Publication Data

Names: Schwartz, Heather E., author.
Title: Showdown : animal groups/ Heather E. Schwartz.
Description: Huntington Beach, CA : Teacher Created Materials, [2018] | Audience: K to Grade 3.
Identifiers: LCCN 2017026877 (print) | LCCN 2017030011 (ebook) | ISBN 9781425853280 (eBook) | ISBN 9781425849542 (pbk.)
Subjects: LCSH: Animals--Habitations--Juvenile literature.
Classification: LCC QL756 (ebook) | LCC QL756 .S39 2018 (print) | DDC 591.56/4--dc23
LC record available at https://lccn.loc.gov/2017026877

Teacher Created Materials
5301 Oceanus Drive
Huntington Beach, CA 92649-1030
http://www.tcmpub.com

ISBN 978-1-4258-4954-2

Many animals share their homes. They live in groups.

Toucans live in the forest.
They live in groups called flocks.

Wolves live in the woodlands.
They live in groups called packs.

Dolphins live in
the ocean.
They live in groups
called **pods**.

Zebras live in the grassland.
They live in groups called **herds**.

Penguins live on the ice.
They live in groups called **colonies**.

Many animals share their homes. They live in groups.

That's just like you and me!

Glossary

colonies

flocks

herds

packs

pod